C13C/4

D0558667

WITHDRAWN
WITHDRAWN

"This book is a really salutary encourag... you think are the obvious ways to frame questions ... possibilities. It is hopeful, realistic and original; what more could you ask from a book about learning to think afresh?"

> *Rowan Williams*
> *Master of Magdalene College, Cambridge;*
> *former Archbishop of Canterbury*

"A stimulating, lateral view of life."

> *Rt Hon Peter Hain MP*
> *Former Leader of the House of Commons;*
> *former Secretary of State for Work and Pensions*

"At a time when the medium has become the message, and transformed the message, it is crucial that we retain the ability to think outside the box and the frame... in other words laterally. This work encourages just that in an entertaining and challenging way."

> *Michael Mansfield QC*
> *Radical Lawyer;*
> *Professor of Law at City University, London*

"This is a book that flips our way of thinking - when we do that we open our eyes to new ways of approaching old and redundant ways of doing. Zoom! - suddenly our perspective shifts and new solutions emerge."

Polly Higgins
Co-founder of the Earth Law Alliance;
author of 'Eradicating Ecocide'

"... a delightful lightness of touch... a wide range of new ways to understand and think about the challenges ahead. It is simply and effectively illustrated, and has insights that will refresh both the expert and someone new to the subject."

Andrew Simms
Fellow of the New Economics Foundation;
author of 'Cancel the Apocalypse'

Framespotting

In a previous century, Laurence and Alison Matthews were university lecturers and statisticians in the oil and transport industries. In this one, they wrote a best-selling book on Chinese characters before turning to epidemiology and the psychology of climate change. Recently Laurence has been Chairman of a climate policy NGO and has given evidence to Select Committees of the UK House of Commons. They live near Hay-on-Wye with far too many books.

By the same authors:
Learning Chinese Characters

Framespotting

Changing how you look at things
changes how you see them

Laurence and Alison Matthews

with illustrations by the authors

BOOKS

Winchester, UK
Washington, USA

First published by iff Books, 2014
iff Books is an imprint of John Hunt Publishing Ltd., Laurel House, Station Approach,
Alresford, Hants, SO24 9JH, UK
office1@jhpbooks.net
www.johnhuntpublishing.com
www.iff-books.com

For distributor details and how to order please visit the 'Ordering' section on our website.

Text copyright: Laurence and Alison Matthews 2013

ISBN: 978 1 78279 689 3

All rights reserved. Except for brief quotations in critical articles or reviews, no part of this
book may be reproduced in any manner without prior written permission from the publishers.

The rights of Laurence and Alison Matthews as authors have been asserted in accordance with
the Copyright, Designs and Patents Act 1988.

A CIP catalogue record for this book is available from the British Library.

Design: Laurence and Alison Matthews

Printed and bound by CPI Group (UK) Ltd, Croydon, CR0 4YY

We operate a distinctive and ethical publishing philosophy in all
areas of our business, from our global network of authors to
production and worldwide distribution.

For Annette and Helen
and the world they will inherit

CONTENTS

Part I: Out of Sight

Part II: Field Trip

Part III: So Now What?

Part I

Out of Sight

Chapter 1

Behind the Scenes

Ever wanted to look behind the scenes at a theatre, a rock concert, or an airport? Of course you have: backstage is a hidden world where exciting things go on, a secret world most of us rarely get to see.

There's a hidden world behind everyday life too. "Framing" operates behind the scenes, affecting how we view things, large and small. In this book we'll show you how to spot when frames are influencing your thinking.

Framespotting can unlock new, more realistic, more effective and *saner* ideas than those we're generally presented with. It can be liberating, too; and you'll see that there's an inspirational story going on all around us, and that you're part of it.

Intrigued? Come and see backstage...

Look At It This Way

How do we get backstage?

Well, the doors are hidden in plain sight. We have to learn to *see* them.

How do we see things? Take society, for example. When we think about changing society's attitude to something, we tend to think of something huge and difficult to turn, like a supertanker.

But we're *not* bits of metal welded together; we're individuals, making decisions independently, but linked together by constant, rapid communication. Perhaps, instead of a supertanker, we're like fish, or a flock of birds: able to change direction in an instant if we need to.

How we see things is important.

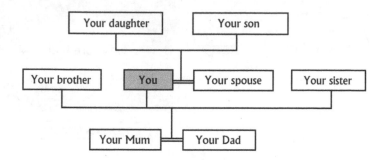

What's Up?

What's wrong with this family tree? Well, nothing. It's not *usual* to put ancestors at the bottom and children at the top, but there's no law against it. Real trees grow upwards, so why not family trees?

It's a choice, like putting north at the top of a map (in medieval Europe, maps often had south or east at the top, rather than north).

So what? Well, these choices can influence our thinking without us being aware of it. For example, we draw organization trees for corporations with the management at the top. This means that managers are to be "looked up to" and so naturally deserve more money. They find this congenial, and commandeer the top floor of office buildings for the same reason. Subconsciously, we think of "up" being better than "down".

Is it just possible that this subtly changes how we see the upper and lower classes, or the global North and South?

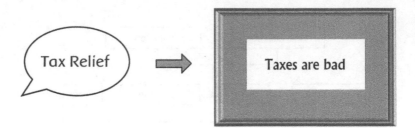

Watch Your Language

The way we see things is influenced by the words we use.

The phrase "tax relief" is like this. It encourages us to think of tax as a burden, or affliction: something we need *relief* from (as opposed to, say, thinking of tax as the membership subscription to a society which provides police, roads and schools).

We seldom think about things in isolation; usually we try and fit them into our view of the world. We try to think about something abstract, like a tax, by thinking about something more tangible or familiar. Like a burden, perhaps.

But looking at things through that "frame" can nudge us into certain attitudes without our realizing it. Burdens are bad, and should be reduced whenever possible. So taxes should be reduced too.

Now maybe taxes should be higher; maybe they should be lower. You decide. But if you're being manipulated into thinking one way rather than the other, this amounts to thought-control. You've been framed.

There's a lot of excitement in Japan about the new K-force minder-robots.

Everyone wants one; it's the latest thing to have.

But there's a snag - K-force can be lethal. K-bots have killed people.

The Latest Thing

What should the Japanese government do about the new K-bots, now that this extremely popular "latest thing" has turned out to be lethal? Should the government ban K-bots?

You'd think so, perhaps. But K-bots bring in taxes and generate jobs. So the government *supports* them. It allows anyone to possess and use K-bots and ignores complaints about them. And the government goes further: people are rarely punished if their K-bot kills someone. The people without K-bots are retreating from previously public areas. The only protection against K-bots seems to be to get one yourself, and so K-bots are spreading, despite the carnage they cause.

What do *you* think about K-bots?

How You See It...

Of course, that story about K-bots was made up.

K-bots don't exist.

Or rather, yes they do. K-bots are cars. K-force is kinetic energy, the force of a heavy object moving at speed: a car can easily have more kinetic energy than a bullet. Cars kill several thousand people *every day* (mostly pedestrians and cyclists). Yet this deadly machine is framed so strongly as an instrument of personal freedom, as a necessity of modern life, that we ignore these deaths. The carnage is accepted as regrettable; there is "no other way".

Such is the power of framing.

*Water? I don't see
any water ...*

*Frames? I don't
see any frames ...*

... How You Don't

Framing works behind the scenes, so that most of us don't realize it's even there. But when you look for it, you'll find framing everywhere.

You do have to look for it, though. The fact that it's everywhere makes it hard to see. The most dangerous thing about framing is that you don't realize it's happening.

A frame is like a camera or computer screen where the view is zoomed in: it shows something in great detail, but it only shows part of the big picture. If you zoomed out, you'd see more of the picture. But perhaps zooming out doesn't occur to you.

We often don't see that we have a zoom control at all.

For small areas, we can use ordinary school geometry to draw maps, mark out fields and so on.

On larger scales, though, things start going wrong, because the earth is really curved. A map of a whole country won't fit onto a flat page without some stretching or squashing. The bigger the area, the worse the problems.

Flat-Earth Thinking

Frames can run deep. They can influence how we understand the world, what we think it *is*, and how it works. For example, is it round or flat?

Nowadays, we know the earth is round and we laugh at people who think the earth is flat. But for much of everyday experience, it *is* flat, more or less. When people didn't travel very far, a flat earth was a rule of thumb which worked fine. It's only when we look at large-scale things, like navigating across oceans, that flat-earth thinking breaks down.

When we accept that the world is round, our whole understanding of the world, our picture of what it is, has changed. Yet that doesn't change things in our everyday lives, like the price of eggs. So a strange thing can happen when you notice a frame, and then choose another frame instead. In one sense, it doesn't change anything; but in another sense, it changes everything.

THE THINKER

To See, or Not to See

At first glance, it seems obvious that the earth is flat. At first glance, it's obvious that management is "above" you somewhere. Or that society is a supertanker.

It's only when you stop to *think* about things that you might begin to wonder. Thinking is hard work, though. It's easier, certainly, to go with the flow; to leave everything zoomed in.

Most of the time, that's fine, obviously. But sometimes it isn't. Sometimes frames can dupe you. But to see this, to think outside the frame, we need to *notice* frames more.

And anyway, aren't you curious? Don't you want to see what happens if you play around a bit with your zoom control?

Chapter 2

Zoom Control

Looking at frames and peeping behind the scenes isn't just interesting. It's also useful to know about, because people can use framing against you. And unless you can see it going on, you won't notice that you're being hoodwinked.

For example, the idea of trying to turn a supertanker is a bit overwhelming. It seems hopeless, or at least a long struggle; you want to give up before you start. And that might just suit *some* people (those who benefit from the current way of doing things) very nicely.

Frames and mental pictures often come with an agenda. They can be used to manipulate and exploit you. And finding your zoom control is the first step towards defending yourself.

What do y<u>ou</u> do?

Sum People...

Laurence started out in life as a mathematician, before working in industry and then writing books.

Now, do you believe that?

Well, it's not true. Nobody starts out in life as a mathematician: Laurence started out in life as a *baby*, just like everyone else.

Oh. That's nit-picking, you say. You knew what that first sentence *meant*. But that sentence sets up a frame: the frame which equates a life with a career.

Think about what that frame leaves out: childhood, family life, the roles of schools and of mothers caring for children. A life is so much more than a career...

Framing can happen easily and innocently, even in everyday conversation.

Lost and Found

It's not always so innocent.

Have you ever noticed how we hear on the news about a factory opening, creating lots of jobs, but then about another one closing, with the loss of lots of jobs?

Well, you win some, you lose some.

But look at the framing here: look at the *words* being used. Corporations happily take the credit for *creating* jobs, as opposed to simply "finding" them. Fair enough. But they don't like to be seen *destroying* jobs; much better if the jobs are simply "lost", a frame which rules out looking for anyone to blame.

Framing is often like this: someone, somewhere is getting away with something, in the sense of not being associated with the consequences.

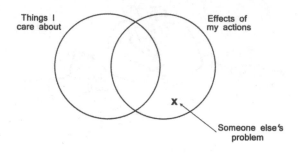

Not My Problem

This applies to stuff being sold to you, too. Cars are framed as freedom machines (think of all those empty roads in car adverts): a frame which emphasizes their advantages and ignores or down-plays their side effects. So you can happily buy one without worrying about any side effects it might have.

Admittedly this framing goes with the grain of our thinking anyway. We don't like having to keep thinking about how our actions might affect other people. It's boring. Restricting.

It'd be much more fun to live in a world where we can just do what we like, wouldn't it?

Of course, that's a make-believe world. Actions *do* have consequences. In a make-believe world they might not, but in the real world they do. But we don't really want to face up to this, and so we lap up a lot of framing about *freedom*.

Cry Freedom

Freedom is a wonderful thing, but it cuts both ways. One person's freedom to pollute erodes the next person's freedom from pollution. It's great to have freedom from being exploited; but what about someone's freedom to exploit *you*: is that great too?

Corporations want "free markets" for example: the freedom to operate without regulation and without having to bother about the pollution or other side effects they cause. But that's denying *us* the freedom to protect ourselves from *them*.

Again there's a general pattern: framing tends to support the freedom of a few people to do certain things, but ignores the consequences, especially for others. The freedom frame often looks at one person's rights, but the next person's rights fall outside the frame and get ignored.

Under the Radar

And now we have a clue to who's *behind* all this framing.

Follow the money. Who benefits? The people who get away with things. By and large the elites, the rich and powerful. The exploiters, rather than the exploited.

How do people exploit you? Well, they can do it by force.

Or they can persuade you to accept what they're doing. They can claim that what they're doing is natural, just the way the world is; if you accept that, you make it easy for them.

Or, best of all, they can arrange things so you don't even notice.

Framing can do this. Framing is a way of limiting the debate, fixing the agenda. It's like a Trojan horse, smuggling in hidden assumptions (and so ruling out alternative ways of looking at things). If the assumptions are hidden, you don't question them because you don't even *see* them.

You'll see them if you zoom out. But who's in charge of *your* zoom control?

Buy, Buy, Buy

Take the frame that sees everything as a market, where you're a consumer rather than a citizen. What assumptions creep in here?

There's the assumption that everything can (and should) be bought or sold; even when talking about education or health.

Anything else? Well, with shopping you can choose not to buy something, or to shop around, but there's no point trying to question what's on offer, or what's *not* on offer. You can choose a red car or a blue car, but not a good bus service.

Corporations don't want us realizing this. Corporations push consumerism, hard. They provide us with all manner of desirable things, but they always want to sell us more. So they sell us the dream of achieving happiness by acquiring ever more stuff. And they try to keep us distracted so that we won't question that message.

They need us insecure and needy, ignorant and compliant.

Corporations want us zoomed in at all times.

Why, Why, Why?

A good way to look for the bigger picture is to keep asking "Why?"

Repeatedly. Don't be satisfied with symptoms; look for root causes. It's like "looking upstream".

Take obesity. Why are so many people obese? Superficially, because of the amount and type of food they consume, and the lack of exercise they take. But ask *why*. Why do people make these decisions? Why is junk food so cheap? Why are people not getting exercise?

A zoomed-in look at car accidents might look at making cars safer. But why was that car there in the first place? Why was it allowed to be going as fast as it was? Why was it even manufactured to be capable of high speeds?

It's easy to see an "obvious" cause of something when you're zoomed in. But is that really the root cause?

The Blame Game

Zooming out usually shares the *blame* out differently.

In the obesity example, the frame which looks at individuals' eating habits *blames* those individuals. Suppose, instead, we started to look at why junk food is available at such low prices, or at deeper causes related to food labeling, food industry regulation and so on. Then we'd start to blame governments and industry lobbying groups. Guess which frame governments and corporations prefer?

The "we are consumers" frame, as well as focusing on markets, also implicitly *blames* us for problems. If we all want cars, or to fly, then climate change is all *our* fault, really. Just like obesity is *our* fault. Now nobody forces anyone to buy junk food, or a car, or a gun; but what happens if we start to think that maybe it's the *producers*, not the consumers, who are the root cause of our problems?

Framing is blaming. And all too often, it's blaming *us*.

Chapter 3

Growing, Growing, Gone

We started off being curious to see what goes on backstage.

But we've been finding that it might be a good idea to know about this stuff for our own good. We're being duped and exploited, and this will carry on until we realize what's going on and stop falling for it.

Actually, it's more serious even than *that*. It's time to look at one frame in particular: growth.

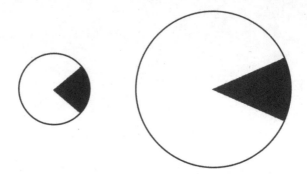

Let Them Eat Cake

Growth is all around us. Politicians and economists are obsessed with it. Growth brings us wealth and jobs, they say. You might think that after all the growth we've had, it's odd that there still seem to be poor people around; but don't worry, we're told, more growth will solve that too.

With a growing cake, everyone feels they can have more in the future. The poor tolerate slack rules for the rich because they dream that they themselves might be rich one day. (Of course, the *really* poor are starving and desperate, but nobody's listening to *them*).

Even if the rich get ever more cake, at least the poor get a few more crumbs than they had before: *some* wealth trickles down, or so the theory goes. (Did you notice that word "down"?)

In other words, growth is used to excuse inequality, even extreme and increasing inequality.

Uh-Oh

Whoa, this is getting a bit political, isn't it? Life's unfair: so what? What's so wrong with growth?

Nothing, for a while. But it can't go on forever. The planet is huge, but it's not infinite: it has limits. We've seen the earth from space; we know this. And recently we've started to hit those limits.

How do we know? Because *planetary problems* are building up: toxic chemicals circle the globe; we're running out of fresh water and turning the oceans acidic; the temperature of our whole planet is rising. There have always been *local* environmental concerns, but these are *planetary* problems, global in scope.

We're creating planet-sized problems for ourselves because *we* are now operating at this scale. Our population, and our actions, are just getting bigger and bigger, and now we're planet-sized. Planetary problems are global because we're outgrowing the planet.

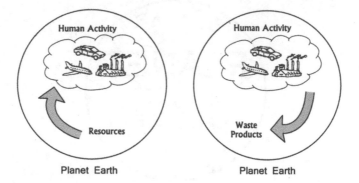

Planet Earth   Planet Earth

Running Out, Filling Up

Planetary problems are hard to see because they're so big: you need to zoom out to see them.

Some are simply a case of running out of resources and raw materials. All sorts of things are getting harder to find; whether it's cheap oil, phosphorus for fertilizers, rare metals for electronics, or supplies of fresh water and land for crops. And we're running out of places to look.

But that's not the worst of it. More insidiously, human activity also generates waste products. This never used to matter too much, but now we're reaching the planet's limits. Landfills, oceans, even the sky: they're all filling up.

Safe places to dump waste are running out too. And it's our waste products that are causing our most urgent planetary problems: toxic pollution, ozone holes and climate change. We're dumping so much that we're overloading our planetary systems.

Life Support

Do such planetary systems matter?

Do they *matter*? Of course they do. Astronauts have a complex life-support system so they can live and breathe in space: complex machines, made up of many parts, keep them alive. Similarly, the water, air, plants and animals of the planet are *our* life-support system. Literally! And, like parts of a machine, they all work together in a complex way; so complex that we may never completely understand it. We're dependent on this machine. Mess with it, or make it malfunction, and we'll die.

And, boy have we been messing with it! We're dangerously close to triggering runaway, unstoppable events that we don't have a hope of controlling. For example, climate change doesn't mean the same world only hotter; it means *climate chaos*: wild and unpredictable swings in rainfall, droughts and storms; a weather system that's gone haywire.

This is Your Life

Now you may feel that saving the planet isn't *your* job...

That's an interesting frame, isn't it, "saving" the planet? It implies an altruistic action, a choice we might make to save something external, like saving an animal.

But this is about saving *us*. The planet is pretty tough, and can take care of itself. But if our life-support systems malfunction, *we'll* be in trouble. We'll *all* be affected, if not directly by droughts, floods and storms then indirectly by rocketing food prices. And by millions of people abandoning countries which become uninhabitable: how will you feel when it's *your* country they all want to come to?

If our crops fail and our food chains collapse, it could mean the end of civilization: worse than the Dark Ages. If humans survive at all, it may be in vastly reduced numbers. Think about that. What do you care about in life? The achievements of your family? Your sports team? It's all at risk. This can destroy everything you hold dear.

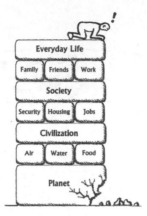

When the Bough Breaks

You're not convinced? There are always people complaining about problems, and claiming that the end is nigh? You don't see any problems in *your* daily experience: on the contrary, everything seems fine. Looking around, there doesn't seem to be any reason to be unduly worried.

But looking around, the earth seems flat.

Your daily experience is the wrong place to look. You need to zoom out. If a foreign army was planning to invade, you wouldn't know by looking for signs in your backyard. Day-to-day life might seem unaffected, until suddenly it's swept away as the foundations collapse.

Later Than You Think

And "suddenly" is an understatement. When something that is growing hits a limit, it happens *fast*. Look around beforehand and there may be tell-tale signs, but on the whole things still seem fine right up until disaster strikes.

There's an old tale of a pond where a lily plant doubles in size every day. It will cover the pond in 30 days. You decide to wait until it covers half the pond before cutting it back. What day will that be? Stop and guess now: is it the 12th day? The 15th?

No. Think about it: the lily plant doubles every day, so it covers half the pond on the 29th day. There's only one day left to save the pond! When this kind of relentless growth approaches a fixed limit like this, it's always later than you think.

Just because something's slow doesn't mean it's not urgent. The time left to take action is silently and invisibly draining away.

A Malignant Growth

Obviously growth can't go on forever on a finite planet, yet we try to ignore this. Why? Because we never seriously question growth. And if something is never questioned, look for a frame.

There's a *lot* of framing around growth. For example, the phrase "economic recovery" suggests that a lack of growth is an illness: something you *recover* from (as opposed to, say, thinking of *growth* as an illness: if you have treatment for cancer and then it starts growing again, you don't call this a *recovery*, you call it a *relapse*). If a resumption of growth is *called* a recovery, who would even *stop to think* whether it might be a bad thing?

We could stop and think about our planetary problems. But frames are so often about steering us *away* from stopping to think.

This can be lethal. Frames are not just interesting to know about; they are not just exploiting you. Some frames are a matter of life and death. Some are threatening our very survival as a species.

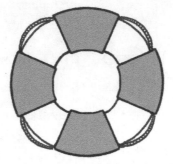

Sharing For Survival

Why is there all this framing of growth? Well, think about what would happen if we accepted that growth had to stop.

If you're adrift at sea in a lifeboat, life is suddenly different. How do you share out the rations? Auction them? Most people would say no; they'd expect to share the food and water out equally. The situation demands a new frame.

If growth stops (as it will have to, that's what a finite planet means) then overall we'll have a steady-state economy. The cake stops getting bigger, so in a post-growth world, the only way for the poor to get more cake is to get some from the rich. When the size of the cake is fixed, people will *demand* more equality. It's only fair.

No wonder the growth frame is defended so strongly. If we stopped to *think* about this, who knows what we might demand?

Playground Thinking

Well we can't have *that*, say the rich to themselves.

So instead, we're all encouraged to believe in a distorted view of the world; to see a make-believe world of no limits. A world of freedom to do anything we want without any bothersome responsibilities. A world where everything's fine and under control. Where growth is good, and anyway essential. Where heroic corporations stride the land, bringing benefits to all: goods, wealth, and jobs. And where *you* don't need to question anything.

The attitude of elites and the largest corporations is like that of spoilt children in a playground. They always want more. They don't want to share their toys. They don't want anyone telling them that they can't just go around hitting the other children; or that they mustn't play with matches that might burn down the whole school. They need a bit of firm parenting.

Chapter 4

Automatic Pilot

So looking behind the scenes started with curiosity, but now we've found some serious framing going on which is blinding us to threats to our very survival.

But surely nobody in their right mind would want that? It can't be true, can it? And if it is, what can we do about it?

Before jumping to answers, it would be as well to see clearly what we're up against.

Firstly, is it true? And if so, is it a cock-up or a conspiracy?

Or are "cock-up" and "conspiracy" just frames too? Are they the only possible ways of looking at the situation?

We need to zoom out a bit more to see the bigger picture.

Playground Addiction

All this framing, especially playground thinking taken together with lavish helpings of propaganda, has the effect of holding us entranced as we sail on towards disaster. It's like being on a drug that has got into the water supply (call it PTD, the Playground Thinking Drug). This powerful drug distorts our view of the world, so that we don't even realize that we're addicted.

PTD is the main reason, deep down, why action to deal with our planetary problems is being blocked, and why our world and all we hold dear is under threat.

Sounds like a conspiracy theory? No, it's not as simple as that. There's no Evil Genius sitting somewhere churning out PTD. But the fact that everyone is hooked on this drug benefits some people more than others; and the main beneficiaries are the rich and powerful.

RoboCorp

Who are the rich and powerful? These days, that description doesn't just apply to people. There are corporations too.

Corporations started out as tools to generate profits, with escape clauses for the owners if things went wrong. Most are useful and beneficial, or at the worst harmless.

But the largest corporations are now like robots that have taken on a life of their own. Set up to maximize profits at all costs, they do just that: they behave in ways that most *people* wouldn't. Many exploit the poor, grab common assets for themselves, cheat, extort and pollute; as much as they can get away with. They want to shrug off *any* restrictions, even on dumping poisons or trashing the planet.

Put like that, it sounds terrible. So it mustn't *be* put like that. Corporations are avid producers of PTD.

The largest corporations have got badly out of hand.

Air Cheapo offers Summer deals

Top Ten Green Buys

The New Russian Arctic

Be the first to see it

What the Papers Say

So are rich elites and large corporations the villains in the story?

Yes, but there are supporting actors too. For a start, the media. TV, radio and newspapers are increasingly locked into pushing PTD. Why? For one thing, they follow subtle (and not so subtle) directives from the corporations that give them advertising revenue. But largely the process is *automatic*, driven by the media's own commercial pressures.

When news is treated as a commodity rather than a vital part of democracy, newsrooms and reporters are cut back to save money. Instead we are fed material supplied by PR agencies in the pay of corporations. Dumbing down drives out serious coverage of the really important issues: some are dramatized to make us fearful, but discussion of root causes is suppressed. News coverage may *seem* thorough, but think of what's systematically being *left out*.

Our media are failing us.

Remember the flat earth? For small areas we can use ordinary school geometry. On larger scales, though, things start going wrong, because the earth is really curved.

Doing geometry for very large areas needs a whole new set of rules. This new geometry for spheres works fine, but only by taking the finite size of the earth into account.

Flat-Earth Economics

Then there are "experts" of all sorts. Economists, especially.

Standard economics is like the school geometry that pretends the earth is flat. It assumes that growth can go on forever, just like the flat plane of school geometry where parallel lines can go on forever. Economists don't usually make a big deal about this; they just assume it without thinking. It's a frame.

But now that we're operating at planetary scale, we need to look at the world as a whole. At this scale we need a new economics which takes into account the finite nature of the planet, just like the geometry we use for spheres. This means changing the frame.

That's fine. We've done it with geometry; we can do it with economics. But most economists don't want to think about this. They like their flat-earth models too much.

Mad economists are just as dangerous as mad scientists ever were.

The Root of All Evil

Banks and the financial industry are keen producers of PTD too, and are avid constructors of frames. There are probably more frames in place around money than anything else (quite apart from the obvious attempts to frame *everything* in monetary terms). A lot of framing tries to hide the process of extracting wealth from us all, and disguise it as services which benefit society.

The whole money system itself is like this. If you borrow money from a bank, the bank doesn't give you some of its own. Instead it creates some in your bank account out of thin air. Then, as well as expecting to be paid back, it demands that you pay interest too. This system gives the banks money for nothing, every day. It also *depends on growth* (to generate the interest). An end to growth would mean an end to this lucrative system. No wonder banks love growth.

We need to reclaim power (and our money!) from the banks.

Follow My Leader

And finally there's government. Politicians largely ignore planetary problems; they too are obsessed with growth. Why?

They'd say that there's no public pressure to get real about planetary problems; the public would even *resist* it. But that's no excuse: the captain of a ship drifting towards the rocks has a responsibility to act, whether or not the passengers are aware of the danger. Why aren't our leaders *leading*? This is an abdication of responsibility on a *colossal* scale. The trouble is, our politicians are hooked on PTD too.

Is it *our* fault? We voted for them, after all. Well, yes, except that in a consumer society where choice is seemingly worshipped, our real choices are actually pretty limited. We can choose a red party or a blue party, but not a government that will lead in the interests of the people, free of pressure from corporations and banks. No wonder we've lost faith in politicians.

We have to reclaim politics.

Go with the Flow

Yes, you say, but that still doesn't explain *why* this all happens. Who's behind it all?

To a first approximation, nobody! That's like asking who's "behind" capitalism. There doesn't *need* to be a conspiracy. We all share responsibility.

We don't *mean* any harm. We're simply channeled into a certain way of doing things. Journalists soon learn what editors want; editors soon learn what advertisers want. People working in corporations soon learn how to behave, what to say. They even pride themselves on knowing how the world works.

And since most people don't even see a problem, that's all it takes. It all happens automatically.

We're not malicious; just unaware. We just think it's the way the world is. We aren't bad, we're just zoomed too far in.

Mad, Quite Mad

Having said that, it's clear that playground thinking is actively encouraged. The oil industry's record on fossil fuels and climate change echoes the tobacco industry's record on smoking and lung cancer. These people know what they're doing.

But again, *why*? You can see why they want more and more (that's just greed), but do they really want to trash the planet?

No, it's simply that they don't *care*. They want ever more money and power, and the removal of restrictions *of any kind*. To achieve this they want everybody hooked on PTD: the general population, people in the media, in education, in government itself. If trashing the planet is a side effect; fine, so be it.

Do they think they're immune? Yes. They're like people in a sinking boat who say, "It's not sinking at my end". They're deluded or, not to put too fine a point on it, insane.

Chapter 5

The Third Story

You may still feel you don't care about planetary problems; you only wanted to look backstage. But beware the framing of planetary problems as environmental issues. Looking through that frame, you tend to think they're something of concern only to some faction called "environmentalists", instead of being crucial to us all. If you view planetary problems as security issues, or as justice issues, does that change how you think about them?

Anyway, if we know about PTD now, can we get on with working out what to *do* about it?

Sorry, no. There's one more layer to go yet.

We're going to dive down to get at some deep stories of who we are; and unless we get that level right, our actions at more small-scale and superficial levels won't be very effective. Frames don't just refer to objects like supertankers; some of the most powerful frames refer to *stories*.

It Was All Going So Well

Let's start by zooming *right* out, and looking at the *really* big picture.

We humans have done pretty well, don't you think? Here we are, flying round the world, communicating globally in the blink of an eye. We understand the workings of the universe, all the way from galaxies down to the intricate mechanisms in our bodies. We store our knowledge in stories, proverbs and books, which form a conversation between humans, across cultures and down the ages.

Along the way we've escaped many of the burdens and risks of the natural world. We've got rid of most of our predators. We build skyscrapers, send rockets into space, and overcome natural limits in lots of other ways too. Of course there are setbacks along the way (and science has given us atom bombs as well as penicillin), but we overcome them one by one, gaining more and more control over nature as we go. Onwards and upwards.

You are here

Masters of the Universe

We live out the stories of our individual lives, and we're always telling each other episodes from these stories. But we also think of ourselves as collectively living out a larger story, the story of the last few decades or centuries for example.

What's the larger backdrop? At the back of our minds there's probably something like the story on the previous page: a story of progress and increasing freedom, a story of control over nature. It's an overarching super-story, in the same way that the super-story of World War 2 forms the backdrop to thousands of smaller stories.

Our "progress" story is a deep story of who we are. It's the story of a climb from obscurity to mastery of creation. It's a story we can be proud of.

And there are no limits to this story. Tomorrow the world, and soon the stars.

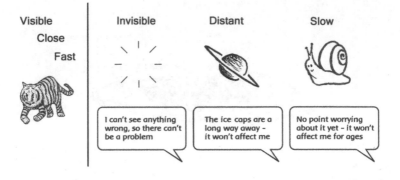

Visible Close Fast	Invisible	Distant	Slow
	I can't see anything wrong, so there can't be a problem	The ice caps are a long way away – it won't affect me	No point worrying about it yet – it won't affect me for ages

Tiger, Tiger

The progress story is a deeply ingrained picture of who we are, and we don't want it questioned.

We see the ice caps melting on TV, but then we try to put it all out of our minds and carry on with life. Why? Because we don't want to be scared, and we don't want to have to change.

This sort of psychological denial is a useful human coping mechanism. It shields us from painful and distressing thoughts. But if something is dangerous we'll usually react quickly, especially if the threat is visible, close and fast: denial wouldn't get a look-in if a tiger was bearing down on you.

Unfortunately, planetary problems aren't like tigers: they're invisible, distant, or slow (sometimes all three). They're just as dangerous, but they don't *seem* so. They bypass our instincts, which respond to threats like tigers, so it's easier for denial to kick in.

Make-Believe

So, problems may be building up as our activities reach planetary scale, but we don't want to know.

The "limitless world" which playground thinking holds out, and which endless growth requires, is a deep frame. This is a psychological issue that we need to confront if we're going to get serious about tackling our planetary problems. But it's hard: we can't really accept, deep down, that our world isn't an infinite playground.

We don't want our progress story threatened. We have an almost childish resistance to any suggestion of limits on our activities: that's why we'd like to pretend that consequences don't exist. We want to live in a world *without* limits. But that's a make-believe world. It's playground thinking.

Make-believe worlds are fun to dream about, and many children live in them. But mature adults have to face reality.

Braking Distance

Meanwhile, however, things are getting *urgent*.

The lily pond showed why we may have less time than we thought. And actually, it's worse than that. Many planetary problems have *tipping points*. A tipping point is like pushing a car over a hill: it starts to gather momentum by itself and becomes difficult or impossible to stop (and think of doing this in fog, because no one is sure exactly where the top of the hill is). With a tipping point, gradual change can seem harmless until, all of a sudden, it's too late.

And it gets worse *still*. We need to act *well before* the tipping points. Think of those Wild West stories where the locomotive is hurtling towards a collapsed bridge. When the driver slams on the brakes, the train doesn't stop instantaneously. There's an achingly long wait as the train slows, sparks flying, getting closer and closer to disaster...

Slamming on the brakes as you hurtle over the edge is too late.

Doom and Gloom

So we need to act urgently, but we're not. Meanwhile, time is running out fast. If you *do* understand all this, it can seem like a horribly inevitable Greek tragedy; it's easy to feel "We're doomed!"

We flip from not doing anything, because there's no problem; to not doing anything, because there's no point.

This "doom" story competes with our progress story. Doom stories fascinate us: we love to watch overambitious people failing. Think Icarus, Babel, Titanic. But do we really want to believe humanity's *overall* story is a doom story?

No. We don't want to give up our progress story, and we hide from the signs that it's breaking down; but we don't want the doom story either. So we try and hide. We fall into denial. And this, as well as being bad for our mental well-being, is leading us towards disaster.

Coming of Age

But are progress and doom the only stories, the only frames?

No. There's a third story.

Growth isn't *bad*: a child grows, and we celebrate this, perhaps by marking progress up a wall. But in adulthood, growth gives way to maintaining health and developing in other ways. Growth, in other words, is something that happens in childhood, and is good only up to a point.

Maturity is when you outgrow growth.

The best story for our times may be a "coming of age" story: if we can realize that a childish fixation on endless growth should give way to a more mature outlook, we will have *grown up*.

And the turning point is here and now.

This is our species' coming of age.

Grown-Up Growth

Although we'll still have pockets of growth as we come to terms with sharing for survival, *overall* growth has to stop. We have to work towards a steady-state world. But this isn't scary or threatening; it's growing up, it's something to aspire to.

And it's only the physical growth of our activities that has to stop (and our economic growth, unless and until we find better ways of unhitching this from physical growth). We can still grow in other ways. The human race can grow in experience, sophistication, knowledge, wisdom, happiness: the things an individual adult acquires throughout life, *without getting any taller*.

So we don't have to totally relinquish our progress story, just transform what it means. Progress from now on will be "inner growth". Physical growth has done its job.

And this may be the end of an era, but it's not the end of the world. When childhood ends, life is just beginning.

Part II

Field Trip

Chapter 6

The Sky's the Limit

In this second part of the book we'll go on a framespotting safari, and then in Part III we'll ask what all this means for us, and what to do about it.

But before we start on our trip, you'll need some inoculations. You'll see why when the destination is announced: we're going to look at *climate change*. Now, quick as a flash, you're probably thinking:

"Oh, that's boring and depressing; can't we do movies instead? And it's so last year; haven't we moved on? Anyway, I've never been any good at science..."

Of *course* you feel these things. You've been framed!

We've chosen climate change because it's one of the most serious and urgent of our planetary problems, and it's an area we know and have worked in. But it's also a great practice ground for uncovering frames: we'll find *seven*.

Not Rocket Science

You might be nervous that the science will be hard to understand.

Come on, you're not stupid. All the science you need is on the next two or three pages, with not a graph or equation in sight.

It's not your fault if you feel this way. As we've seen, many large corporations are doing everything in their power to distract you from thinking about things like climate change, because they want you to buy their products and not look too closely at what they're up to. In particular, they don't want you to think or do anything that would lead to any restrictions on their activities. In other words, corporations are, by and large, strongly against you looking at serious issues such as climate change.

So it's all framed as being difficult and best left to experts. Don't fall for it.

High Stakes

But didn't you also read somewhere that the science isn't *certain*?

You're right: there's overwhelming evidence about the basics, but we're still learning about the details. But think what that means: we have to make decisions without knowing everything, which means looking at *risks*.

Now, how we deal with risks depends on what's at stake. When the stakes are low, as with trying out a new product or a new route to work, you're happy to take risks. When the stakes rise, such as wondering what might happen if your house burnt down, you take out insurance. With very high stakes, you try to avoid risks altogether: you'd probably refuse to play Russian roulette, even if insurance was available!

With planetary life-support systems we shouldn't be taking *any* risks we can possibly avoid. Claiming that we can delay acting until we're sure of all the details is like playing Russian roulette with our children's lives.

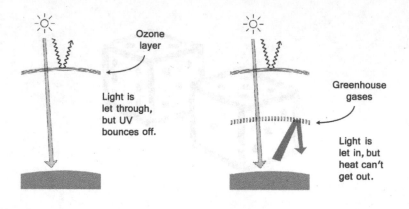

Ozone layer

Light is let through, but UV bounces off.

Greenhouse gases

Light is let in, but heat can't get out.

Duvets in the Sky

The ozone hole was our first warning. We found that chemicals from refrigerators and aerosol cans were eating into the ozone layer in the sky, which shields us from cancer-causing ultraviolet radiation. Oops. Suddenly, here was a planet-wide system which we almost crashed before we even realized what was going on.

Lower down in the sky, *greenhouse gases* are part of another planetary system. Like the glass in a greenhouse, they trap heat. They let sunlight *in*, which warms the earth, but they stop the heat getting *out* again. By dumping more and more greenhouse gases into the air, we're causing the planet to overheat, like a fever patient in a hospital. And the last thing you do for a fever patient is to pile on duvets to heat them up further!

Fevers can be lethal: the body's systems break down. The earth's systems are complex, with all sorts of components (oceans, jet streams, ice caps, monsoons), and we really don't want *them* breaking down.

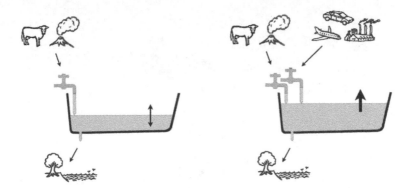

A Greenhouse in your Bath

The best-known greenhouse gas is carbon dioxide, or CO_2.

CO_2 in the air is like water in a bathtub. As CO_2 is added to the air (like a tap adding water to the bathtub) and removed again (like a plughole removing water), the level varies up and down. The amount of CO_2 in the air (measured in units called ppm) has varied in the range between 180 and 280 as the ice ages have come and gone.

But recently, burning fossil fuels (coal, oil and gas) has added lots of extra CO_2. It's as if we've turned on another tap, more and more each year. Together, the two taps are adding more than the plughole can cope with, and the water level has been rising. The CO_2 level is now around 400, far higher than it's been for a million years, and it's still going up.

You can think of climate tipping points as the bathtub overflowing: a new set of problems altogether. To prevent this, we need to stop turning the second tap on. In fact we need to turn it almost all the way *off*, to give the plughole a chance.

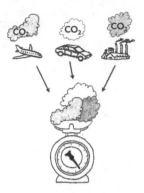

The Scales of the Problem

Turning off that second tap means reducing our carbon emissions or "carbon footprint". In turn, that means leaving a lot of fossil fuel in the ground. We simply can't afford to burn it all: the CO_2 would wreck our life-support system.

The problem isn't a scarcity of fossil fuels; it's that the sky is filling up with greenhouse gases. The sky really is the limit.

Now the climate system doesn't care where the CO_2 comes from; it's the total that matters. If you imagine weighing all the CO_2 emissions from the world each year on a gigantic set of scales, then all that matters is the total: that's what we have to reduce.

This may sound blindingly obvious, but it doesn't stop some countries claiming to have reduced emissions when all they've done is to move them around. (For example, you can emit less CO_2 in manufacturing if you buy Chinese goods, but the same emissions then just happen in China instead).

It is no use saying,
"We are doing our best".

You have got to succeed in
doing what is necessary.

Does It Add Up?

When countries offer to cut their emissions by various amounts by various dates, does this help?

It depends. Numbers matter. It's like bridging a stream with a plank: you need a plank long enough to reach across. It's not enough to find a short plank and say, "It's a step in the right direction." The size of the response needs to match the size of the problem.

And we need to move *fast*: all the time we're turning the second tap off, it'll still be pouring out water; the two taps together will still be outpacing the plughole; and *the water level will still be rising* (remember the Wild West train). To get the CO_2 back down from today's level (around 400) to a "safe" level (around 350) we need cuts in emissions that are *deep* and *fast*.

It's no good reducing by "a bit" or "what we can get agreement to". We need "whatever it takes to get to 350", pure and simple.

Chapter 7

Name That Frame

Now let's look at our response to this situation. Along the way, we'll look out for frames and hidden assumptions, and we'll draw attention to the frames as they crop up.

Spotting frames is all very well, but remember that the real benefit from doing this is that we can then go on to see what we might have been missing *outside* those frames.

We'll be looking out for better ideas, where "better" means more helpful and more effective; and often this means more zoomed out.

Nero Tolerance

We need to respond fast, and on an appropriate scale, to avert climate chaos and all that goes with it. Are we doing this?

No.

When it comes to tackling climate change, our governments are being feeble. There are some measures in place (emissions trading schemes, carbon taxes) but they are hopelessly inadequate when measured against the only question that matters, namely: "Will this stop runaway climate change before it's too late?" Leave it to governments and it seems they will fiddle while the earth burns.

At this point, a visiting alien might observe that this is strange. Surely the first duty of governments is to protect their people. Why aren't they dealing with this problem, *urgently*?

I Will If You Will

One problem is that governments find it hard to work together. They may try to cooperate, but at the same time they're all frantically competing with each other. Each makes excuses and tries to leave the hard work to others. The USA might treat the situation as a game against an opponent and hope that China, say, will act first. But China may think along the same lines...

It's easy to get trapped by this game-playing. But that's not a reason to give up. That would be like discovering gravity and then giving up trying to build bridges or aircraft.

If the rules of the game are stacked against us, then we need to change the game. Only then will we make progress.

FRAME: We're so used to thinking about the world as if it's made up of countries that we often don't realize that this is a frame. Are there other ways of looking at the world? We'll come back to this frame later.

Adapting to Apocalypse

Something governments *do* seem to find it easier to talk about is "adapting" to climate change. Their response to sea-level rise or increasingly severe storms is to build flood barriers.

> **FRAME:** Governments try hard to frame policies on climate change as *responses* to it. The only problem with climate change is how to live with it, they imply. This frame stops us asking whose *fault* it is, or what can be done to *stop* it.

Adaptation is like rearranging deck-chairs on a sinking ship instead of fixing the hole in the hull. It doesn't *tackle* climate change, which will just get worse and worse, overwhelming each defensive measure.

Some effects of climate change are already inevitable, and so some adaptation will be needed. But to rely on adaptation *alone* simply isn't tackling the problem.

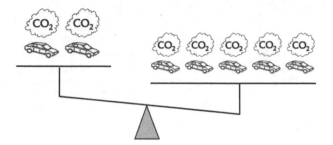

Efficiency Bites Back

If we can't leave it to governments, what about industry? How about being more efficient?

Well, all those green developments are good, don't you think? We're getting greener cars, greener planes, even greener insurance policies! But do greener cars and planes bring down carbon emissions? Obviously, yes...

FRAME: *NO!* There's a frame here focusing on the *wrong thing*. The "efficiency" frame looks at emissions *per car*, but ignores the *number of cars*.

If greener cars simply encourage more of us to buy them and to travel further, we could easily emit more CO_2, not less. Greener planes may emit less CO_2 *per passenger-mile*, but the growth in passenger-miles is outpacing this, so we have *more* CO_2 again.

Trusting industry to be more efficient isn't the answer either.

Smoke and Mirrors

Oh dear. But we're clever, aren't we? Surely we'll think of something. Already there are claims that technology can ride to the rescue. For example, *geo-engineering* (ideas like blocking sunlight using space mirrors or spraying particles into the sky) promises to halt or reverse global warming directly.

Well, yes, we'll need all our wits to get through this crisis, and it would be silly not to develop back-up plans. But geo-engineering only tackles one symptom (temperature), leaving other symptoms (like acidic oceans) untouched. And if a Chinese space mirror cools Russia, then what? Geo-engineering *alone* is not the answer.

FRAME: Framing climate change as a technical problem invites us to think it doesn't involve *us*. Do you think we should bet all our lives and futures on geo-engineering? Or is it a distraction, an excuse not to stir ourselves politically and tackle playground thinking?

> Today we're talking to Sandalman, from the Ecoliving movement. You believe that the only way we're going to stop global warming is by making big changes to the way we live, right?

> Yes. We'll all have to cut down on car travel, really cut down on flying, eat more local food, insulate houses better, give up some of our stuff...

> So you want us to go back to living in caves?

Total Rewind

If leaving it to industry or government isn't working, perhaps it's down to *us*. Many people feel this and have made changes in their lives, big or small, hoping to set an example to others.

The trouble is, nobody likes the sound of "giving up" things (even if greener lives often make people healthier and happier). It's all too easy to see Sandalman as someone who wants to "spoil our fun" or "destroy our jobs / economy / way of life".

In the long term Sandalman is right, but in our current world it's unrealistic to ask everyone to change overnight: it's too big a jump for most of us. So Sandalman's ideas are easy to ignore or to parody.

FRAME: This "green means austerity" frame is widespread. But is it true? We gave up slavery, but we're immeasurably richer now than the slave owners ever were. (And try "having fun" when your life-support system fails).

**Reduce your
Carbon footprint
NOW!**

Doing Your Bit

You probably don't want to live like Sandalman, but perhaps you're prepared to make *some* changes. If so, what should you do, as an individual?

"Do Your Bit," you're told. "Change your light bulbs, turn down your thermostat, insulate your loft... it's all down to you."

This is a *trap!*

Because, of course, not everyone *does* do their bit. We need a system enforced by law. Otherwise efforts by some individuals will be pointless, undermined by the actions of others. Let's face it, hardly any of us would pay tax if it were voluntary; but we do pay (and hence we get roads, libraries, the police) because we know that by law everyone else has to pay too.

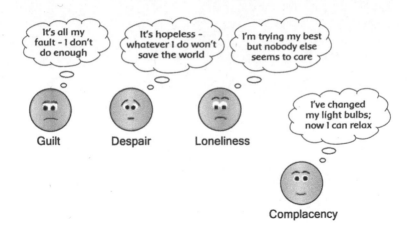

Guilty Atoms

The psychology of Do Your Bit is interesting, don't you think? Unless you, as an individual, do your bit, any resulting catastrophe will be *your* fault. So you feel guilty for not doing enough, even though you can see many others doing even less, or nothing at all.

You despair, because the mismatch between the huge problem and any one person's tiny contribution is all too obvious. And it feels very lonely, trying to save the world on your own.

On the other hand, Doing Your Bit gives you a good excuse to sit back, feeling that at least you've done *something*.

FRAME: Notice how getting you to fixate on your individual efforts distracts you from seeing that there's no policy in place to tackle the problem *as a whole*.

Doing Your Bit isn't the answer either. So what *is*?

Chapter 8

All Change

Looking at the frames we've uncovered, it's no wonder that we're not taking climate change seriously. We pretend it isn't there, or we pin our hopes on technology, politicians or token attempts to do our bit, and hope that it'll all go away. Meanwhile, consciously or not, many of us are worried or fearful: we know these simplistic efforts aren't working; we've got stuck.

But that's not the end of the story. Deep down, we probably knew it couldn't be that easy. We have to be smarter. But how?

Well, our tour has shown us frames which normally sit there unnoticed: six so far, and another one coming up. Once we've *seen* them, the way is open to look for ideas those frames might have been blocking.

Let's start by taking the problem seriously: ignore the "adapting to it" and "efficiency" frames; accept that numbers matter. How can we move that pointer on the scales to where we need it to be?

Chasing Emissions

Suppose, first, that we're looking to reduce a country's carbon footprint (we'll look at the world presently). We want to cap, or limit, the country's carbon emissions at a safe level. Clearly, we need to regulate the factories, power stations, vehicles and homes responsible for those emissions; that's how "Cap and Trade" schemes work, and it's how carbon rationing would work too. But things get complicated, bureaucratic and expensive very quickly. And up to now, carbon-trading systems have been full of loopholes. Carbon rationing would be complicated too, and nobody likes the idea of rationing –

Stop! Stop. Look at that word "clearly".

FRAME: There's a frame here, which focuses attention on where the emissions *take place*. Let's try something else.

Upstream, Downstream

If you had a garden hose connected to a sprinkler, then to save water you wouldn't try to block up the sprinkler holes; you'd simply turn off the tap a bit. Now, think of fossil fuels going into the hose from the tap, and emissions coming out of the sprinkler holes. Aha! By controlling the *fossil fuels* coming *into* the system (the tap), we automatically control the *emissions* created further down the line (the sprinkler holes). This is capping emissions "upstream".

How does it work? The simplest way is for the government to set the desired CO_2 limit, and then to sell CO_2 permits for that amount. This determines "the amount of water to be allowed out of the tap". The fossil fuel suppliers (the oil, gas and coal corporations) buy the permits. They can then sell the amount of fossil fuel covered by the permits they have bought.

The sprinkler holes take care of themselves because the tap does the work. Simple.

Think upstream.

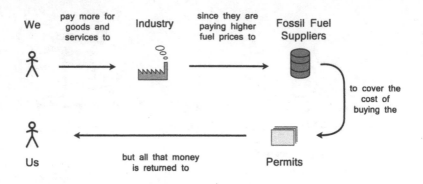

Carbon Cashback

With such a system there's no need for us to look at our individual emissions or work out our carbon footprints, no need for anybody to be checking up on us. We can largely forget the system's even there, and go about our lives.

So does the system affect us at all? Yes. The price of carbon-intensive goods goes up: the fossil fuel suppliers have to recoup the cost of buying the permits somehow, and they pass the cost on to us.

Oh. Not good. But wait: who gets the money from selling the permits? Why not *us*? Let's share it out: every adult gets an equal share, say. After all, it's our country, so it's our system and it's our money. This "carbon cashback" money we get then covers the increase in fuel prices. In fact, if you have a low or middling carbon footprint, you'll make a *profit*: your cashback payments will outstrip the price rises.

It's just like a strong carbon tax, strong enough to do the job; except that *we* get the money, not the government.

Fair
Simple
Cheap
Positive
Effective
Fast

Cheap and Cheerful

This general approach, capping emissions upstream and sharing the benefits equally, is called *Cap and Share* (not to be confused with Cap and Trade). It has two parts: the Cap or emissions limit, which is no-nonsense (there are no loopholes and meeting this emissions target is guaranteed); and the Share, which is fair (it has to be, or people won't accept the Cap).

It's simple and cheap, because only a small number of fossil fuel corporations are involved. Yet it does the job.

And it has a positive psychology to it. The "green means austerity" frame always tries to portray tackling climate change as a *cost*. But Cap and Share solves the problem without costing us anything. Fossil fuel prices do go up, which they have to if we're going to change our ways, but the money is all returned to us!

Think positive.

All Together, Now

Carbon cashback rewards you for having a low carbon footprint. But of course, in a more important sense *everyone's* a winner: we all get to avoid climate chaos for ourselves and our children.

This is a shared, *collective* way forward. It avoids the guilt and divisiveness of relying on people to do their bit voluntarily. We own the system together, and all contribute to it and benefit from it. In other words, we are rejecting the framing of climate change as something that is down to individuals to solve.

After all, we reject this framing in other contexts. Think about immigration. We don't get everybody to patrol a bit of the border, do we? No, we expect the government to do this on our behalf: that's what governments are for. It's the same with carbon emissions.

We're all in the same boat here.

Think collectively.

My emissions are all within the cap, and now I'm paying for them too, so leave me alone!

At last we're back on a survival course, thank goodness...

I'm making money out of this

Something for Everyone

There are a variety of arguments for taking action along these lines. Firstly, like it or not, we have to take effective action somehow: it's essential to our survival. Millions of lives are at stake, and what we'll get out of it is a future for our children.

Sadly, this answer isn't enough for everyone. But if they benefit in a more tangible way too, then persuading them to support the right thing is much easier. So in this tangible sense, who'll benefit and how?

The majority, including almost all poorer people, will benefit directly and financially. The rich, with high carbon footprints, won't; but they'll get something else. They can say, with justification, that their emissions are now at least part of an overall plan, and moreover a part which they are paying for.

People can do the right thing for a variety of reasons. We don't have to rely on guilt, or altruism.

Think pragmatically.

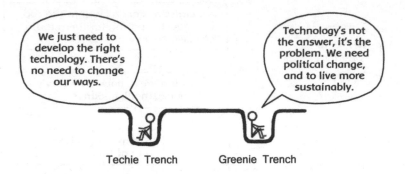

Techie Trench   Greenie Trench

In the Trenches

But won't we need technology too?

Yes: technology is one of our major strengths. But it seems to evoke one of two opposed, entrenched positions: most people either think technology is the answer to everything, or hate it and think it's the root of all our troubles. Neither of these positions is very helpful.

We need technology: getting the right policies in place may set the right course, but we also need technology to adapt *in practice* to lower carbon lifestyles.

But conversely, technology *alone* won't save the day without changes in policies and priorities.

The fact is, we'll need both technology *and* to change our values and how we live. We need our cleverness *and* our wisdom.

 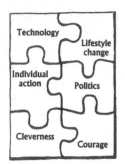

Takes Two to Tango

We also need local *as well as* global approaches. Many people are focusing on local issues, with an emphasis on communities, local food production and so on. But this must work hand in hand with global thinking: without global policies to tie everything together, the global situation will deteriorate and overwhelm any local efforts.

With most of these either-or questions, the answer is "both". We need to be practical, but still hold on to our idealism. We'll need individual change *as well as* collective action; and ingenuity *as well as* courage.

It's not one or the other; we need both.

Think inclusively.

Tomorrow, the World

And finally: we've talked about how to cap a country's emissions, but we need to cap emissions on a *worldwide* scale. How can we do this?

The obvious approach is to set national caps using a formula agreed by all nations, adding up to a global cap (a level that will "get the world back to 350"). Each nation then operates its own scheme to achieve its agreed national cap. Easy!

If only! The rub, of course, is agreeing the formula. We've seen how easily "I Will if You Will" and other game-playing can cripple negotiations to agree such an approach.

So, look again at that frame that sees the world as being made up of countries. Is that how the climate system sees it? If you look at the earth from space, do you see countries?

Forget countries and look at the world.

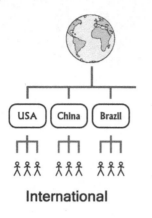
International

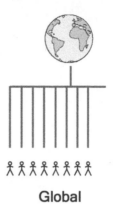
Global

International Warming

What about a single, worldwide scheme for the planet as a whole? Maybe a World Climate Organization operates a worldwide system of permits for fossil fuel corporations, with the resulting money returned to the (world) population.

A *global* approach like this, as opposed to an *international* one, shortcuts all the national posturing and game-playing. It embodies global equity. It emphasizes the unity of humankind.

Of course, it won't be quite that simple. National governments will resist global solutions, because each government wants control over what happens in its own country. In most aspects of life, of course, they should have that control. But global issues need global solutions, and global emergencies need global action.

It's global warming, not international warming.

Think globally.

Part III

So Now What?

Chapter 9

I Have a Dream

OK, forget about climate change now; back to frames.

The field trip in Part II aimed to give you a feel for framespotting. Whatever the topic, zooming out can lead to lateral thinking; it encourages long-term thinking, and it can reveal new perspectives and powerful, deep stories.

We can't just leave it at that, though. Having ideas won't get them implemented; knowing about a story won't make it happen. If you want to explore what we can *do* about some of the frames in our lives, read on.

In Part I we dived down to get to some deep stories, and now it's time to come back up to the surface. But let's come up slowly, like a scuba diver avoiding the bends. Before we jump to tactics, we need to get better strategies and stories in place at a deep level. We may be against playground thinking, but what are we *for*?

And talking of stories, we'll find that stories can tell us a few things along the way.

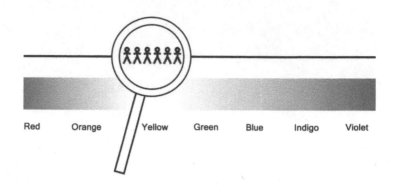

Red   Orange   Yellow   Green   Blue   Indigo   Violet

Rainbow Vision

If the human beings alive today lined up along a rainbow spectrum, with the poorest at the red end and the richest at the violet, where would *you* be? Maybe you don't think of yourself as rich. But if you have a car, for example, you're already up at the indigo end of the scale.

Consumerism and the celebrity culture are zoomed in on the violet end of the spectrum. However rich you are, consumerism encourages you always to be looking enviously at people who are even richer. But those people at the extreme violet end of the spectrum are the ones causing most of the problems. They're the ones most avidly wrecking the planet. Don't envy them; *stop* them.

And notice that we tend to look at lots of issues through this frame of comparing our lot with that of others. That's a game where nobody can gain unless someone else loses. Is that what life is *really* all about?

The Common Good

Like players in a football team, we're all individuals but we're also part of something that we do together.

Our road system and energy grid are *common* assets. Effective media, justice and police: these are common assets too, hard won down the ages. We should be proud of them. And a planetary life-support system is the ultimate common asset.

Playground thinking, on the other hand, is largely about selfishness, about grabbing or exploiting common assets for private gain. Corporations want to use road systems or an educated workforce, but they don't want to pay for them.

Myths and stories are in no doubt about selfishness. Heroes are selfless, in touch with others and fighting for them; villains are always selfish and cut off from others. Which side are *you* on?

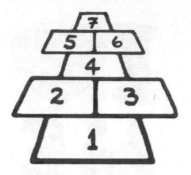

One Step at a Time

Concern for common assets means protecting them; stopping them being eroded or wrecked. But this doesn't mean a centralized bureaucracy prying into all aspects of our lives; still less a change in human nature. All we need is a few changes in the rules.

We've done this many times before. The abolition of slavery was just a change in the law; it didn't change human nature or solve all our problems. But it did mean that everything now operated in a framework where at least slavery was (mostly) behind us and we could move on to the next thing.

And the next thing at the moment is to set up frameworks to take care of our planetary problems, so that we can survive and thrive in the future.

Get a Grip

Changing the rules is a political task. But we're not talking revolution here: we don't need to overthrow capitalism or abandon democracy.

We need more democracy, not less. If corporations are currently playing one country off against another, then we need strong democratic control to rein them in again.

We probably still need capitalism too, but we need it to operate within a grown-up set of rules which acknowledges planetary realities.

And we need our political leaders to get a grip and to chart this course on behalf of humanity, rather than allowing a few people with megaphones to drag us off course.

Look Back in Wonder

Picture the future world of a few decades' time, where grown-up thinking prevails. Technology has moved on; there are many wonders. But the main wonder is that people are more awake, saner, more at ease; more aware of a common purpose; more *alive*.

This is a future to look forward to. From this standpoint we'll probably look back on many of our current values with something like amused pity, just as today we look back on aspects of the past (such as fashion) with a mixture of nostalgia, embarrassment and disbelief. To our children, the frantic binge when we were caught up in playground thinking will seem old fashioned and pitiable. It was heady for a time, but ultimately lethal and insane. It was a chapter in our story, but better was to come.

Future Imperfect

Again, picture this future world. OK, it isn't paradise. Much of day-to-day life is unchanged; there's still inequality, injustice and strife as well as the good things in life.

But obscene inequality is declining. Cars seem like yesterday's technology. Each country and each region is a little less insecure, a little more self-sufficient. Corporations no longer make us feel quite so insecure, fearful and despairing. Planetary problems are being tackled; we have every reason to hope that we're on a survival path. Our future is secured as a foundation for all our other hopes and plans. Who'd go back?

An adult life stretches ahead for our species. Do we want this brighter future? Now's the time to decide: to face our challenges or to hide; to bring about a prosperous future world, or to let it slip away.

Chapter 10

From Here to Maturity

Moving beyond playground thinking is a *huge* challenge. Can we do it? Perhaps trying to kick the addiction to PTD is hopeless from the start. Perhaps the PTD culture is too entrenched.

Here's the doom story again: in theory we *could* save ourselves, but in practice we aren't clever enough; or perhaps we're too clever by half.

There's often a low point in stories where all seems lost. But this is the point when the hero decides to fight back. *We* have to fight back too; there's everything, almost literally, to play for. But we'll need to give it all we've got.

So what's the plan? Well, of course, there's no simple Answer To Everything. But there are some pointers...

And what does it *feel* like, suddenly to find ourselves in this position?

Leadership Challenge

The first reaction might be anger. Why have "they" got us into this mess? Politicians, for example, are abdicating all responsibility for setting a survival course. Rage may be justified, but politicians are our representatives as much as our leaders. And there *is* a lack of public protest; so maybe we're getting the politicians we deserve.

It really *is* down to us. But we need to change politicians, not light bulbs. We need to act like citizens again: building grassroots support and pressure, besieging politicians and above all spreading the word to get others on board too.

To encourage grown-up politicians, *we* have to be more grown-up. If a few politicians try to move in the right direction, we need to give them support and to resist opponents (and snipers in the media) who try to ridicule them. We have to encourage politicians to stand up to playground thinking.

We need political leadership; but we have to help it to emerge.

Common Cause

Looking at obesity, child poverty or anything else, we need to zoom out, look upstream, trace symptoms back to their root causes. But as we do this, several things happen which can make us uncomfortable. Seeing the bigger picture may be exhilarating but can be unnerving. We get into unfamiliar territory. And it all gets bigger, and the vested interests more powerful. Lots of bad news, then.

The good news is that we also discover more and more allies, who can bring undreamed-of strength to our fight. Environmentalists concerned with our planetary problems are enemies of PTD; but so are the justice, peace and equality movements that are concerned with the common good or the public interest. So are many faith groups, and groups fighting on specific issues like reform of the banking industry or curtailing advertising aimed at children.

We all need to join together. My enemy's enemy is my friend.

First they ignore you
Then they laugh at you
Then they fight you
Then you win

Worth Fighting For

In stories, insanity is worse than ignorance: at the centre of the web, the Bond villain is always mad (quite mad), and must be fought.

It's scary? Yes. We're vaguely aware of this, and that's one reason why we try to hide in denial.

But being born is scary; so is leaving home. We may not *want* to wake up from a cosy dream to a starker reality, but waking up is the only way to be fully alive, to hold our heads high and look the world in the eye. PTD will try to play on fear. Choose instead to focus on life.

Remember what we're fighting for: a world where we can face our children and grandchildren with a clear conscience. How do you want to be remembered? There are people who will tell you not to think about any of this, to get on with consuming. But don't stand for being short-changed like that. This is your birthright, your life-story they want to take from you. Don't let them.

Paper Tigers

And, again, there's good news too.

PTD may seem powerful, but it's not half as powerful as you might think. Much of its apparent power is an illusion produced by PTD itself.

Many corporations are like paper tigers, fierce-looking but fragile. If we suddenly stop buying their stuff or using their services, they'll be running scared. *We're* powerful too.

In stories, monsters always have a blind spot. Bring monsters from the shadows into the light, and you see their weakness.

This is why spreading the word is so important. PTD only fools those who are addicted to it, who don't see its effects. See and be free.

Get Real

Time for a reality-check? Cynics will say that these ideas are all very well, but we have to live in the real world.

Isn't that an interesting frame? By "the real world" they mean political reality. But the *real* real world isn't political, it's *physical*: it has atoms and temperatures. Politics is an invention, much less fixed and "real" than political "realists" would have you believe.

Physical reality trumps political reality every time. Political wheeler-dealing and human nature can't be ignored, but the ultimate realism is to accept the laws of physics.

In a way, this is *good* news: if we're fighting to change the status quo of growth, the universe is on our side, since the finite size of the planet will *force* us to change, one way or another. Cynics clinging to "political reality" are fighting a losing battle.

Season Finale

We're living through an exciting time, a time that will be seen as a major turning point in our story. It's been called names like the Great Turning or Great Transition. Such language may appear grandiose, but it's not some dreamy new-age thinking that's driving this; it's pure, stark logic and the laws of physics. We're outgrowing the planet, and it's happening *now*. We have to realize that we can't carry on as we are. If we don't want collapse, then we'd better get this transition started.

It's a climax, but there have been others in the past. Our human story is a *big* story, which in turn is part of the even grander story of the universe itself. Phases of our story are almost like seasons in a long-running TV series, and now we're at a season finale, with the inevitable end-of-season climax.

We want to make sure that there'll be a *next* season.

You Ain't Seen Nothing Yet

It may take a crisis. We seem to be a species that needs a crisis to make us act. But a crisis is coming soon, to a planet near you.

We're getting close to the edge, and time is short. If everyone wakes up to this situation, it may very well provoke a crisis. A "great awakening", a psychological tipping point, can be very sudden and dramatic, and there may be widespread fear that we've left it too late.

But the thing about a crisis is that it forces action. All sorts of things become possible that were previously inconceivable. Even governments can act decisively, if pushed by people and by vested interests who suddenly wake up to a danger to themselves. Such a response might only be provoked by a crisis threatening our very existence, but when that response comes it will be something to see.

Touch and Go

So, are you optimistic? Can we do it? What grounds are there for hope? Not a simplistic hope which just waits for miracles, but a realistic, honest hope: a hope that we can overcome our problems if we face them head on.

We can look around for signs, to try and guess the odds. There *are* some hopeful signs, although they may seem few and far between. It'll be a close-run thing, for sure. But as the climax approaches in a story, it always looks impossible that the hero will triumph, or even survive. But they do.

Will *we* make it? The only honest answer is: nobody knows. In the meantime we have to stop thinking too much about the odds and start acting to improve them.

Chapter 11

Interesting Times

So, as our story of humanity's coming of age reaches its climax, what can each of us do, as individuals, to play our part in this great story?

Oh; you never wanted to be in a story? Tough: it's going on around you. But you *do* get to choose the part you'll play. You can be an active player; or a small contributor; or you might be a background character who ignores the big picture and just gets on with life. But you have a role of some sort, simply because you're alive at this moment on the earth.

Living the Dream

The deepest, most fundamental thing you can do is easy, and doesn't cost anything. It's simply to see yourself, not only as a player in the day-to-day events of your everyday life, but also as part of the greater story. Use your zoom control. It changes nothing, but it changes everything.

Live your values in your own life; be alive to the choices that come to you. Make your choices, not (just) to save the world, but because it feels right; because you're in tune with the new, adult future.

We should be checking each choice in life against the test, "Is this grown-up thinking, or playground thinking?" With the big picture at the back of your mind, the lure of PTD and the temptations of runaway consumerism are easier to resist.

Wake-Up Call

The next thing is easy, too. There may be no Grand Plan, but for sure the first step is to wake people up. Write to your political representatives, but don't stop there. We *all* need to wake up to the crisis, before we get to the edge.

Spread the word, tell the story. Point out playground thinking. Argue back, gently to begin with: at the shops, in the pub, at your place of worship, with your fellow workers. It's the way we've changed attitudes to drink-driving and countless other things, big and small.

Draw attention to frames. Zoom out, and help others to do the same.

Look at what playground thinking wants us to do, and what it says about us. We're better than that.

The Long and Winding Road

This is a long haul. The transition may take decades; it may feel like a "long emergency" rather than a climax. Meanwhile, you have a life to lead, and it'll be exhausting to be thinking about planetary problems all the time.

Good advice is to be a "part-time crusader": enjoy life, but keep the big picture in mind. When and where possible, strike a blow in the long struggle. Live your life as if your future depended on it.

We can each set an example in our own lives. But seriously, we also need to push for *collective* action: to demand changes in the rules which we can then all embrace as team players. Many organizations are already pushing. Choose one and join it, especially one that collaborates with others.

You aren't alone: you don't have to do it all. If you are facing in the right direction, you'll find others alongside you. Support each other. You *can* make a difference; but that's you plural, not you singular.

Do the Right Thing

And finally, hold in your mind a picture of what it means to be fully human. What's *right*.

At root, our planetary problems are a moral challenge, not a technical, economic or even political one. Ensuring our own comfort at a terrible cost to our children's future is not worthy of us as moral beings. Slavery was wrong; apartheid was wrong; so is this. We can't claim to be people of integrity unless we stop the powerful from destroying creation.

And your own part in this? Well, that is (of course) up to you. We all have different concerns, and are able to make different contributions. Dramatic roles may fall to only a few of us, but even small voices can help ripples spread in today's connected age. In times of flux and change, small nudges can make huge differences.

You may have a big part to play, or a modest one, but it's a part in a huge, wonderful story. Play it.

Pictorial Recap

We think in boxes, or frames, zoomed in so that we don't notice the bigger picture. And often we don't realize this is happening. We need to zoom out.

Framing can be innocent, but it's often deliberate. We've been framed; we get blamed. Zooming *right* out, we're outgrowing the planet and making dangerous problems for ourselves.

We urgently need to get serious about these problems. And we *can:* we're capable of wonders; we can think laterally; we're ingenious and resourceful.

But instead there's confusion, game-playing, calls to "Do Your Bit". It's all pretty ineffective. Why? Well, these are hard and deceptive problems. But there's something else holding us back.

We're entranced by playground thinking: there are no limits, growth is good and everything's fine. It's like a drug, PTD, that holds our media, academics, politicians - and us - mesmerized.

If we *did* take planetary problems seriously, or question growth, then we'd start to look at fairness; and at who's to blame. That would affect profits. The PTD pushers can't allow that, even at the cost of our survival. There's insanity here.

We have to escape this addiction. This is a political fight, and the first step is to wake everyone up. We need to be both clever and wise. We need to look at basic values, forge alliances. We need to do what is right.

The prize is a grown-up world, where we will have outgrown physical growth. This is our species' coming of age. Here and now. And you are part of it.

Acknowledgements

Blaise Pascal famously wrote, "I have made this longer than usual because I have not had time to make it shorter." We make no apology for having taken the time, however tempting it was on many occasions to write a "normal" book!

But to get there, through the months and months of edits and re-writes, required the assistance of some very special people. For taking the time and being prepared to say what they really thought (good and bad), huge thanks go to Martin Adfield, Denise and Malcolm Darbyshire, Brian Davey, Barbara and John Eastwood, William Fisher, John Funnell, June and Paul Graves, Jane and John Neville, Sarah and Dave Morgan, Geoff Petty, Dil Roworth and Liz Singh.

In addition there were those who encouraged, commented, listened and helped in diverse ways. Thanks to Dave Bennett, Annie Bowdler, Anna Bradley, Thomas Cullis, Larry Fogg, Andy Fryers, Peter Gunn-Wilkinson, Mark Landis, Sue Lee, Geoff Rothwell, Judy Seymour and Richard Shaw. Special mention must, as always, go to our chief cheerleader and advisor on all things American, Judy Landis.

And finally Annette and Helen: for being helpful, encouraging and downright wonderful people, our love and thanks go to you both.

A number of illustrations were inspired by great (in some cases iconic) cartoons and artwork by the following artists. Apologies to: James Steidl (page 14), Ron Leishman (page 19), Joe Shuster and Ethan Van Sciver (page 70), Alfred Leete (page 72), Ellis Nadler (page 76), cg4tv.com (page 94), Rex May (page 98), Roy Delgado (page 103) and Start-Art (front cover).

Notes

Several of the ideas in this book have been drawn from other sources, as listed below. More details, plus extra material and suggestions for further reading, can be found on the website www.framespotting.com that accompanies this book.

We first heard the supertanker / fish analogy in a talk prepared by David Wasdell for the All Party Parliamentary Climate Change Group of the UK House of Commons.

The "tax relief" frame comes from the book *Don't Think of an Elephant* by George Lakoff, and the lily-pond tale is quoted in *The Limits to Growth* by Donella Meadows and others.

The bathtub metaphor is used by Aubrey Meyer of the Global Commons Institute, famous for "Contraction and Convergence", and the hose analogy appears in the book *Climate Solutions* by Peter Barnes, which outlines "Cap and Dividend" (the US version of Cap and Share). See www.capandshare.org for more details, and see www.350.org for more about 350 ppm.

The realization that conventional economics denies any limits is widespread (except among conventional economists), but not the direct analogy with flat-earth thinking. However we have recently discovered the same idea in the 1988 book *Economics for a Round World* by Charles A. Pierce.

Producers' role in consumption was highlighted by Rupert Read on the "Green Words Workshop" blog.

We haven't seen the "coming of age of the species" metaphor expressed elsewhere as explicitly as we have done, but the idea of growth as a phase like childhood appears in many places, for example *The Growth Illusion* by Richard Douthwaite and *Cancel the Apocalypse* by Andrew Simms.

The idea of a "part-time crusader" is due to Edward Abbey.

BOOKS

Iff Books is interested in ideas and reasoning. It publishes material on science, philosophy and law. Iff Books aims to work with authors and titles that augment our understanding of the human condition, society and civilisation, and the world or universe in which we live.

FOL

SEP 2 2 2023